Anderson Kokiti Basseto

Internal combustion engines

Anderson Kokiti Basseto

Internal combustion engines

Comparing the Otto Cycle and the Diesel Cycle by Analyzing Efficiency and Pollutant Emissions

ScienciaScripts

Imprint
Any brand names and product names mentioned in this book are subject to trademark, brand or patent protection and are trademarks or registered trademarks of their respective holders. The use of brand names, product names, common names, trade names, product descriptions etc. even without a particular marking in this work is in no way to be construed to mean that such names may be regarded as unrestricted in respect of trademark and brand protection legislation and could thus be used by anyone.

Cover image: www.ingimage.com

This book is a translation from the original published under ISBN 978-620-2-18321-5.

Publisher:
Sciencia Scripts
is a trademark of
Dodo Books Indian Ocean Ltd. and OmniScriptum S.R.L publishing group

120 High Road, East Finchley, London, N2 9ED, United Kingdom
Str. Armeneasca 28/1, office 1, Chisinau MD-2012, Republic of Moldova, Europe
Printed at: see last page
ISBN: 978-620-7-44630-8

DEDICATORY

I dedicate this work to my family, especially my parents, for everything they represent in my life.

ACKNOWLEDGMENTS

First of all, I thank God for everything.

To my parents and family who have always helped me and given me the strength to complete this work in the best possible way.

To all the teachers who contributed to my education, especially the course coordinator, Prof. Dr. Daniel Salinas, my advisor, Prof. Dr.[3] Edneia Santos de Oliveira, for her attention and support in the construction of this work and also to the teachers who helped in some way.

To my colleagues and friends who have always been by my side in times of difficulty and joy.

Finally, to all those who in some way contributed to my education and helped make this work happen.

The greatest reward for a man's work is not
what he gets out of it, but what he becomes
out of it.

(John Ruskin)

BASSETO, A. K. **Internal combustion engines: comparison between otto cycle and diesel cycle analyzing performance and pollutant emissions**. Foz do Iguaçu: UDC, 2016.

SUMMARY

Despite the development of engines using renewable fuels, the current market still uses internal combustion engines, which in turn use non-renewable fuels (fossil fuels). The internal combustion engines used in land vehicles can be classified into two main groups: the Otto cycle and the Diesel cycle. Because they burn fossil fuels, these engines are relatively closely linked to environmental pollution and oil shortages. The internal combustion engine is one of the biggest contributors to air pollution and the depletion of crude oil reserves. Given these important issues, this study will present a comparison between the Otto cycle engine and the diesel cycle, using two vehicles with the same characteristics, an S10 (Flex) pick-up truck and an S10 diesel pick-up truck. The methodology of this research consisted of experimental tests and tests carried out on a dynamometer and gas analyzer, where the results of performance (torque and power), fuel consumption, and emissions were compared between the fuels, alcohol, gasoline, and diesel. In terms of performance results, the fuel that gave the engine the highest power values was alcohol, and diesel had the best torque. In the efficiency test, gasoline had the best result. In general, this work highlighted alcohol fuel, both for its good performance results and from an environmental point of view, as it comes from biomass, its source being considered renewable and part of the CO_2 emissions captured in the plants that give rise to the fuel.

Keywords: fuels, engines, pollutants.

TABLE OF CONTENTS

1 INTRODUCTION

Currently the market is saturated with vehicles, and issues such as fuel efficiency, alternative energies and vehicle emissions have become of great importance, due to the fact that these vehicles are contributing to environmental degradation, so issues related to the environment and sustainable development have been increasingly discussed.

The World Organization of the Automotive Industry (OICA, 2011) states that the global vehicle fleet has already reached one billion land vehicles. In Brazil alone, the fleet has already exceeded 30 million units and the number of vehicle licenses in the year (2013 and 2014) was over 5.26 million, which represents a growth of over 15% in the fleet in just two years (ANFAVEA, 2015). This relative increase in land vehicles using the internal combustion engine has led to an increase in land vehicle congestion in cities around the world, which is generating an increase in polluting gas emissions in ever higher proportions, causing environmental imbalances such as global warming. In addition to this increase in pollutants, there is the scarcity of non-renewable sources which, according to Pompelli *et al* (2011), the rate at which these sources are being consumed means that they cannot be renewed for use again, which means that they are expected to run out in a relatively short time.

The internal combustion engine is one of the main contributors to air pollution and the depletion of crude oil reserves (MARTINS, 2013). According to this statement, reducing pollutant emissions and improving the performance of internal combustion engines is of great importance to society, and most land vehicles are still powered by internal combustion engines (BOSCH, 2005). The purpose of these engines is to transform thermal energy into mechanical energy, which enables land vehicles to move through the transmission. The internal combustion engines used in land vehicles can be classified into two main groups: the Otto cycle and the Diesel cycle (BOSCH, 2005).

Despite the development of vehicles powered by engines using renewable fuels, the current Brazilian market still uses vehicles powered by internal combustion engines using non-renewable fuels (POMPELLI *et al,* 2011).

Because combustion comes from fossil fuels, engines are relatively closely linked to environmental pollution and the scarcity of oil. Studying the cycles of internal combustion engines is of great importance, because through data and tests you can compare the relationship between the Otto cycle engine and the diesel cycle, determining which would have the best efficiency with the lowest emission of polluting gases.

1.1 GENERAL OBJECTIVE

To make a comparative analysis of the performance of a pickup truck (S10) with an Otto cycle engine in relation to another with a diesel cycle engine, measuring the parameters of fuel efficiency and pollutant emissions.

1.2 SPECIFIC OBJECTIVES

- Evaluate performance tests.
- Evaluating pollutant gas emissions tests.
- Evaluate performance tests.

2 THEORETICAL FRAMEWORK

The purpose of the theoretical background is to present the definitions and concepts from the literature on what will be presented in this final course work. This chapter discusses the main themes linked to the research, based on the scientific literature. The topics covered are linked to internal combustion engines, fuels, engine performance, efficiency and pollutant emissions.

2.1   FUNDAMENTALS OF INTERNAL COMBUSTION ENGINES

According to Torres (2010), internal combustion engines are "thermal machines that convert the heat energy generated by combustion into mechanical energy capable of producing movement." Bosch (2005) adds that internal combustion engines generate energy by converting the chemical energy contained in fuels into heat, which is converted by combustion. The heat generated by combustion produces mechanical work.

2.1.1 History of Internal Combustion Engines

The idea of engines came about in 1652 with Father Hautefoille, who proposed building an engine using the expansive force of the gases from the combustion of gunpowder in a closed cylinder (TORRES, 2010).

Although nothing was recorded on the subject, in 1680 Huygens proposed an idea similar to the priest's, which would be an engine that would use gunpowder as fuel and would work with a cylinder and piston (VARELA, 2010).

After seven years (1687), Dénis Papim succeeded in developing and describing the working principle of a piston steam engine, and in 1767, James Watts developed a steam engine with a cylinder cooling system (BOSCH, 2005).

According to Martins (2013), the first patent for an internal combustion engine only appeared in 1794, which was obtained by Robert Steet, in which the engine would consist of two horizontal cylinders, which would be connected by a chain drive and would produce the energy that would be used to move the power cylinder. It would be the first internal combustion engine in which liquid fuel would be placed directly in the cylinder, but the inventor did not build the proposed engine.

In 1797, B. Thompson and Count Rumford observed the equivalence between heat and work during the construction of a cannon. This increased the possibilities for engine calculations. In 1801, Phillip Leben obtained a patent for a combustion engine, which worked on the principle of expanding gases from the combustion of a mixture of air and ignited gas (VARELLA, 2010).

After two decades, W. Cecil was the first to develop a combustion engine that worked successfully, in which he worked with a mixture of air and hydrogen. Following this invention, Jean Etienne Lenoir, following Cecil's principles, began a project to build an engine in 1852, but it wasn't until 1858 that the first fixed gas explosion engine was launched, which he patented in 1860 (TILLMANN, 2013).

From the origin of this engine, they began to think about the possibility of transforming rectilinear movement into rotational movement, so in 1863, Lenoir built a tricycle that used a gas engine made from shale or light oil (shale or tar) that was vaporized by a primitive carburetor of only 1.5 HP. It was noted that the ignition mechanism (carburetor) was of great importance for the start-up of internal combustion engines, due to the difficulties encountered in starting the engine (TORRES, 2010).

The process of running the engine took place by means of the gas being compressed inside the cylinder, then combustion was generated by means of an electric spark, there being no mixture between fuel and air, because only the fuel (the gas hilha) was compressed in the cylinder (TILLMANN, 2013).

Lenoir, the inventor of the tricycle, failed to understand the importance of the mixture of fuel and air in the combustion process, as the air provided an increase in heat production due to the increase in the amount of oxygen. Despite the success of the tricycle, which traveled all over Europe and even won the Argenteuil Grand Prix (Paris - Joinville-leponte motor race), it was never marketed (VARELLA, 2010).

In 1854, the first 2-stroke combustion engine was built by Dugald Clerk, but the engine was only presented in 1881. After presenting the engine, the German Gottlieb Daimler envisioned reducing the size of the 2-stroke engine by introducing hot-spot ignition, which made it possible to build automobiles years later (TILLMANN, 2013).

Three years after building the 2-stroke engine (1857), Barsanti and Matteuci built a free piston engine that used the expansion of combustion gases to propel the piston vertically upwards, and when the piston descended by the action of gravity, it drove a ratchet, which in turn drove a shaft (MARTINS, 2013).

This engine was marketed by Otto and Langen until 1867, and in 1862 Beau de Rochás proposed and patented the working principles of 4-stroke internal combustion engines with pistons, which due to their characteristics were highly efficient, although he was unsuccessful in applying his theories, i.e. he never managed to build his engine. In 1876, Nikolaus August Otto ended up building the engine and calling it the silent Otto, after having independently invented the same cycle described by Beau de Rochás (VARELLA, 2010).

Approximately 15 years later, the compression-ignition engine was developed by Rudolf Diesel in 1893, which was called the Diesel cycle engine (TORRES, 2010).

2.1.2 Classification of Internal Combustion Engines

According to Martins (2013), internal combustion engines can be classified in several ways: in terms of their use, the movement of the piston, the phases of the work cycles, the number of cylinders, the arrangement of the cylinders, the properties of the intake gases and one of the main ones in terms of ignition, which is the otto cycle where ignition occurs by a spark, heat or spark produced by a spark plug. And the diesel cycle where ignition occurs by compression between fuel and air.

2.2 OTTO CYCLE

The Otto cycle originated even before Nikolaus August Otto, as the first step was taken by Beau de Rochas in 1862 when he announced the "four-stroke" engine, which was unsuccessful (TILLMANN, 2013).

Figure 1 shows the Beau de Rochas engine cycle diagram, where cycle A corresponds to the theoretical cycle and cycle B to the real cycle (JR, MARTINELLI, 2008).

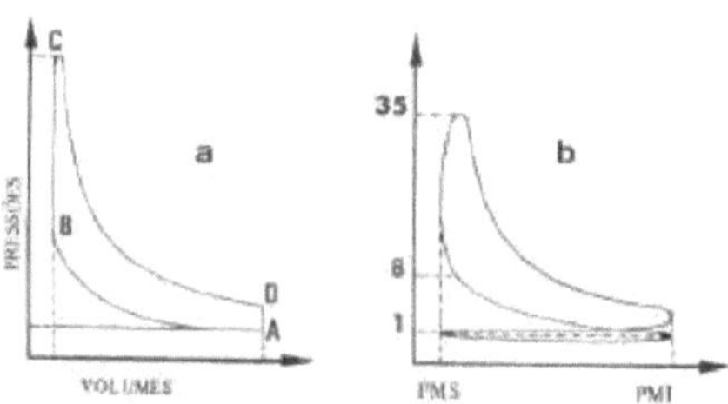

Figure 1: Beau's Rock Cycle (a) Theoretical Cycle (b) Real Cycle.
Source: JR, Martinelli (2008).

Although Beau de Rochas was unsuccessful with his engine. The German Nikolaus August Otto continued and, in 1876, ended up building the engine, which gave rise to the name "Otto Cycle" (VARELLA, 2010).

According to Martins (2013), the Otto cycle engine works as follows:

The 1st stage (intake) is where a mixture of air and fuel is admitted.

2nd time (compression) the mixture is compressed.

3° Time (explosion) combustion occurs, where it is initiated by a spark that is produced by a candle.

The 4th stage (exhaust) is where the gases produced by combustion are expelled.

In the 1st stage an air/fuel mixture is admitted, which according to Martins (2013) is obtained when the carburetor or electronic injection (more modern engines) injects fuel and the intake admits oxygen contained in the air, forming the air/fuel mixture. For each fuel there is a ratio of air to be admitted and fuel to be injected, in the case of gasoline it is 14.8:1, 14.8 parts air to 1 part gasoline. For alcohol it's 9.0:1 - 9.0 parts air to 1 part alcohol.

Figure 2 shows a schematic of the Otto cycle engine of an S10 (FLEX) pickup truck, which uses both gasoline and alcohol as fuel.

Figure 2: Otto cycle engine (S10-Flex)
Source: Prepared by the author

2.3 DIESEL CYCLE

Diesel cycle engines originated in 1892 when engineer Rudolf Diesel published a booklet in Berlin entitled "Theory and construction of a rational heat engine", in which he set out his ideas for putting Said Carnot's cycle into practice. It was known that the first engine of this cycle could not be realized. Rudolf Diesel abandoned the cycle due to the dangers it presented due to its high compression and replaced it with a simpler cycle, known as the "Diesel cycle", in which he began building his first engine in 1897, which was nothing more than a single-cylinder with a diameter of 250 mm, stroke of 400 mm and consumption of 247g of fuel per horse\hour, which developed 20HP at 172rpm and a thermal efficiency of 26.2% (TORRES, 2010).

According to Martins (2013), the engine developed by Rudolf Diesel works on a four-stroke system and basically has two major differences from a gasoline engine: Firstly, the engine sucks in and compresses only air and secondly, the injection system doses, distributes and sprays the fuel into the cylinders, after which the fuel ignites when it comes into contact with the air, which is strongly heated by the compression. It uses a compression ratio of approximately 19:1.

Figure 3 shows the engine cycle diagram, where diagram A corresponds to the theoretical one and diagram B to the real one (JR, MARTINELLI, 2008).

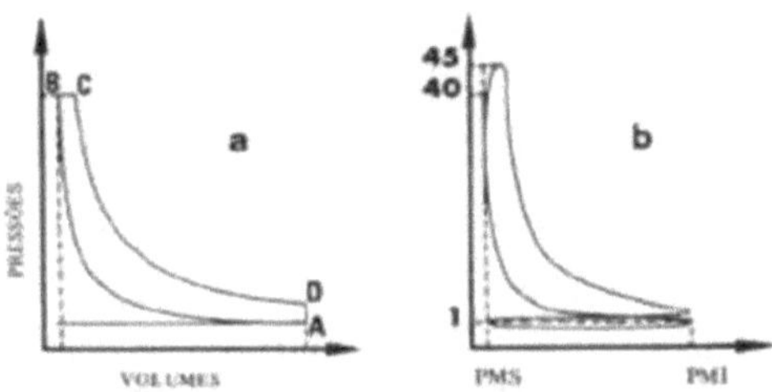

Figure 3: Diagrams of Rudolf Diesel's cycle (a) Theoretical Cycle (b) Real Cycle.
Source: JR, Martinelli (2008).

According to Martins (2013), the cylinder in this cycle is filled and emptied at atmospheric pressure, because:

AB = adiabatic compression of the pure air sucked in before.

BC = combustion at constant pressure.

CD = adiabatic expansion.

DA = sudden drop in pressure.

The first phase is adiabatic compression, where pure air is sucked in and compressed to a temperature sufficient to ignite the injected fuel, while in the second phase (isobaric compression) combustion begins and takes place at constant pressure, when the volume increases and the expansion of the gases compensates for the drop in pressure due to the increase in volume. The third phase (adiabatic expansion) is where the expansion takes place without heat being exchanged with the cylinder walls and the last phase (low pressure) is when the exhaust is abruptly opened, producing a rapid drop in pressure while the piston oscillates in neutral (constant volume) (TILLMANN, 2013).

The main difference between the Otto and Diesel cycles is that in the Otto cycle combustion occurs due to a spark produced by a spark plug, and in the diesel cycle combustion occurs due to compression between fuel and air (BOSCH, 2005).

Figure 4 shows a diesel cycle engine used in an S10 pickup truck.

Figure 4: Diesel Cycle Engine **Source:** Prepared by the author

2.4 CHARACTERISTICS OF THE FUELS USED IN OTTO AND DIESEL ENGINES

Each fuel has different compositions and properties, providing different combustion reactions, performance, efficiency and emissions. These fuel properties are relatively closely linked to the characteristics of the engine's combustion process. Important properties include: calorific value, fuel anti-knock index, vapor pressure, ignition temperature, density, activation energy and enthalpy of formation (ALBAHRI et al., 2003).

2.4.1 Resistance to "Detonation" of Fuels

Detonation", popularly known as "knocking", is related to the fuel's ability to be compressed (air/fuel mixture) without spontaneous combustion. These properties are related to a fuel's octane rating (BOSCH, 2005).

Based on the octane number of the fuel used, engines can be designed with higher compression ratios and earlier ignition advance points, which can lead to improvements in performance parameters and the engine's thermal efficiency (MARTINS, 2013).

In spark-ignition engines (Otto cycle), the octane number indicates the anti-knock characteristic of the fuel, which strongly depends on the type of hydrocarbon present in the fuel (TILLMANN, 2013).

The two methods normally used to determine the octane number are the RON (Research Octane Number) and MON (Motor Octane Number) methods (ALBAHRI et al., 2003).

2.4.2 Fuels Used in Otto Cycle Engines

2.4.2.1 Petrol

Gasoline is a fuel with a complex mixture, used in Otto cycle engines, which contains more than a hundred different chemical compounds. Its chemical formula varies according to the refining conditions and the type of oil it comes from. Worldwide, gasoline is characterized by its octane rating. In Brazil, gasolines are marketed with octane ratings that are within international standards (FERREIRA, 2003).

One of the factors that influences the octane rating of gasoline is the chemical structure of the hydrocarbons present in fuels. Where the introduction of a linear chain double bond to make an olefin has a great effect on increasing the RON of the fuel, while in MON the proportion of the increase is much smaller (OWEN and COLEY, 1995).

In the 1970s, the Brazilian government began using ethanol blends in pure Brazilian gasoline, the main motivating factor for this change being the 1973 oil crisis (ANFAVEA, 2015). As a result, the gasoline currently used commercially in Brazil has a composition that varies between 25% anhydrous ethanol and 75% pure gasoline (type A), giving rise to type C gasoline, which is called regular gasoline in Brazil.

2.4.2.2 Alcohol

Alcohol is a fuel that can be used in Otto cycle engines and has some advantages and disadvantages over gasoline. One of the advantages is that it is a renewable fuel, reducing dependence on oil consumption (CHO E HE, 2008).

According to Owen and Coley (1995), alcohol has excellent octane qualities, which means it burns cleaner and has good properties for use in spark ignition engines.

Alcohol has a higher anti-knock index than gasoline. This property can be taken advantage of in an engine if it is well designed, in which parameters such as the engine's compression ratio, ignition advance time and stoichiometric ratio can be varied, generating an increase in pressure inside the chamber, resulting in higher torque and power values (BOSCH, 2005).

One of the disadvantages of using alcohol is that in low-temperature conditions, which are usually critical at temperatures below 10° C, it makes it difficult to drive and start the engine due to its low volatility compared to gasoline (MARTINS, 2013).

Generally, to minimize this problem, gasoline is injected when the vehicle is started, in some cases the fuel is heated before starting and ethanol is mixed with gasoline. In Brazil, the type of alcohol sold at gas stations is hydrated ethanol, which contains around 7% water by volume (TILLMANN, 2013).

2.4.2.3 Natural Gas

Natural gas is an alternative fuel used in Otto cycle engines, which provides a cleaner burn due to its completely closed fuel system, so evaporative emissions are practically non-existent (ABIANEH et al., 2008).

According to Owen and Coley (1995), because natural gas is found in the earth, it contains varying amounts of non-methane hydrocarbons, H_2S, CO_2, water vapor, nitrogen, helium, argon and other gases. These gases and hydrocarbons generally require treatment to prevent damage from corrosion and condensation.

Another characteristic of natural gas is that it is a fuel with a wide flammability range and a high octane number, which allows the engine to work with mixtures that are poorer than the stoichiometric condition, making it advantageous in some situations of engine use (ABIANEH et al., 2008).

According to Cho and He (2008), it is possible to apply higher compression ratios in poorer mixtures compared to the stoichiometric mixture, because the fuel is more resistant to detonation.

2.4.2.4 Changes in the Engine's Operating Parameters Depending on the Fuel

As the Otto cycle engine uses several fuels, in which these fuels have different properties to each other, adjustments to the engine's control maps are necessary for operation with each of the fuels (HEYWOOD, 1988).

The first parameter is the fuel dosage. Each fuel requires a different fuel injection timing map. For each fuel there is a mixture that fits the stoichiometric conditions. These mixtures are due to the percentage of oxygen contained in the exhaust gases, in which the sensor informs whether the mixture is "rich", excess fuel, or "poor", short of fuel, and each fuel contains its appropriate percentage (CHO AND HE, 2008).

These mixtures are regulated by a regulator in the fuel line. The regulator is an electronic system that controls the opening time of the fuel injection valves in order to measure and proportionalize the amount of air admitted. Another parameter that needs to be corrected is the number of spark ignition points in the engine. To form the ignition maps, the values vary according to the fuel used and the engine conditions, such as load, speed and temperature. There is an ignition point for each engine operating speed, which optimizes combustion pressure and improves engine performance (HEYWOOD, 1988).

2.4.3 Fuels Used in Diesel Cycle Engines

2.4.3.1 Diesel oil

Diesel oil is a fuel used in diesel cycle engines, which is derived from the distillation of petroleum and selected according to its ignition and flow characteristics. Consisting basically of saturated hydrocarbons, diesel oil is made up mainly of carbon atoms, hydrogen and, in low concentrations, sulphur, nitrogen and oxygen (ALEME, 2011).

Diesel oil is obtained from crude oil, which requires several processes, including cracking and fractional distillation. These processes are necessary because they limit the correction of the specific mass of diesel oil as a final product, which is a determining factor in altering the characteristics such as performance and gas emissions of an internal combustion engine (VALENTE, 2007).

In order to meet the sustainability requirements demanded by society and the laws in force, Petrobrás has refined different types of oil, reducing their sulphur content, such as S500, S50 and S10 diesel, which refer to the ppm sulphur index in diesel (PETROBRÁS, 2012).

In today's market, with the introduction of EURO 5 technology engines, only S10 diesel oil is recommended, i.e. with a total sulphur content of no more than 10 mg kg^{-1} or 10 ppm (PETROBRÁS, 2012).

2.4.3.2 Biodiesel

Biodiesel is a mixture of oilseeds that is used as a sustainable alternative fuel that is widely used in diesel engines. In addition to oilseeds, vegetable oil is also used in its virgin state or from waste oils from frying processes, animal oils, etc (KNOTHE *et al.,* 1997).

Around 100 years ago, Rudolf Diesel, the inventor of the diesel cycle, proposed the use of biodiesel from oilseeds, in this specific case peanut oil, a project that opened the door to research that has been carried out to this day (MOTHÉ *et al.,* 2005).

From a technical point of view, biodiesel has a very important characteristic in terms of the number of cetanes present, obtaining a greater energy gain in combustion than commercial diesel oil, i.e. less ignition delay (FERNANDES *et al.,* 2008).

As far as the environment is concerned, the use of biodiesel is very important, as it can reduce SO_2 by up to 90% as a result of automobile exhaust gases (SOCCOL *et al.,* 2005).

2.5 CALCULATION CONCEPTS FOR ENGINE PERFORMANCE AND EFFICIENCY

2.5.1 Acceleration

According to Halliday *et al.* (2008) acceleration is when the velocity of a particle varies, because for each movement along an axis, the average velocity in a time interval (Δt), is given by Equation (1):

$$a_{med} = \frac{v_f - v_i}{t_f - t_i} = \frac{\Delta_v}{\Delta_t} \tag{1}$$

In other words, the acceleration of a particle at any moment is the rate at which the velocity is varying in an instant. The unit of acceleration in the SI is the meter per second squared, (m s- 2). Other units are used in the form of (length-time^{-2}) (RAMALHO *et al.,* 2007).

2.5.2 Force

According to Newton's second law "the resultant force acting on a body is equal to the product of the mass of the body and its acceleration" (HALLIDAY *et al.*, 2008).

This concept in a mathematical way results in Equation 2

$$\overrightarrow{F_{res}}=m\vec{a} \tag{2}$$

This equation means that when the force acting on a body is zero, the acceleration of the body will be $a = 0$. If the body is at rest, it tends to remain at rest, and if it is moving, it continues to move at a constant speed. The unit of force in the SI is Newton, which tells us that $1N= (1\ kg)\ (1m\ s^{-2})=1\ kg\ m\ s^{-2}$, in other systems of units the force can be given differently (RAMALHO *et al.*, 2007).

To solve problems involving Newton's second law, we need to draw a free-body diagram, where the only body shown is the one for which we are adding the forces. These forces acting on the body will be represented by arrows with the origin at the point, and the acceleration of the body will be shown by another arrow (accompanied by its specific symbology). A coordinate system is also generally used (JEWETT AND SERWAY, 2012).

2.5.3 Power

Power is a derived quantity, where it is obtained from the quotient of energy and a unit of time. The so-called "mechanical work". time^{-1} as can be seen in Equation (3) (HALLIDAY *et al.*, 2008).

$$P_{méd}=\frac{W}{\Delta t} \tag{3}$$

This power consists of nothing less than force multiplied by speed, which is given by Equation (4) (RAMALHO *et al.*, 2007).

$$P=\vec{F}.\vec{v} \tag{4}$$

Although power is derived from a vector quantity (force). Mechanical work is considered to be an absolute quantity, as it is the force that acts in the direction of the body's displacement, so there is no need to understand its direction in order to understand its meaning (JEWETT AND SERWAY, 2012).

Considering that "mechanical work" means the same thing as mechanical energy, we can call mechanical work mechanical energy. In this way, power is the amount of energy generated per unit of time, and is therefore a scalar quantity, which does not require knowledge of direction to understand (HALLIDAY *et al.,* 2008).

The value of "mechanical work" mechanical energy (in joules) divided by the unit of time in seconds results in power, represented in SI by the unit watt (W): 1 W=1 j s^{1} (RAMALHO *et al.,* 2007).

This was just a guide for illustration, because at the time of the Industrial Revolution it was difficult to visualize the working capacity (power) of a machine. As (many) horses were used to drive machines. So James Watt (who lent his surname to the SI measure of power) illustrated the measure of power in a different way, in which he estimated the capacity of a horse to vertically lift a mass of 33,000 pounds at a speed of 1 foot per minute. This measure of power became known as horse-power (HP) (JEWETT AND SERWAY, 2012).

From a numerical point of view, "horsepower" is very close to hp, but comes from different origins. The hp is considered to be a mass of 75 kg (thus "doing" a force of 75 kgf or 735.4 newtons) lifted vertically by one meter in one second (RAMALHO *et al.,* 2008).

2.5.1 Torque

According to Halliday *et al.* (2008) torque is represented by the force exerted on a lever arm coupled to a shaft to cause it to rotate. It sounds strange in words, but you can visualize it mathematically (5).

$$\tau = r.\ F \tag{5}$$

It would be like a wheel wrench placed on a perfectly tightened bolt. The force it makes on the wheel wrench is multiplied by the length of the wrench, which is represented on the bolt. If this bolt offers no resistance to torsional movement, a zero torque will be displayed; if the resistance is high, a high torque will be displayed (JEWETT AND SERWAY, 2012).

Torque is expressed in newtons x meter (N-m) according to SI, although the no less correct unit meter x kilogram-force (m-kgf) is in common use, where 1 m-kgf = 9.806 N-m (RAMALHO *et al.*, 2007).

In an internal combustion engine, torque is proportional to the lever arm, which in the case of a vehicle is the central distance between the trunnion and the journal at the moment the piston explodes inside the combustion chamber (Figure 5) (RIBAS, 2003).

Crankshaft

Moentcs : cylindrical surface (indicated by the arrowhead) on which the connecting rods of the 4 pistons rest. This is the crank that bears the force resulting from the explosion of the air-fuel mixture that occurs one at a time inside the cylinders.

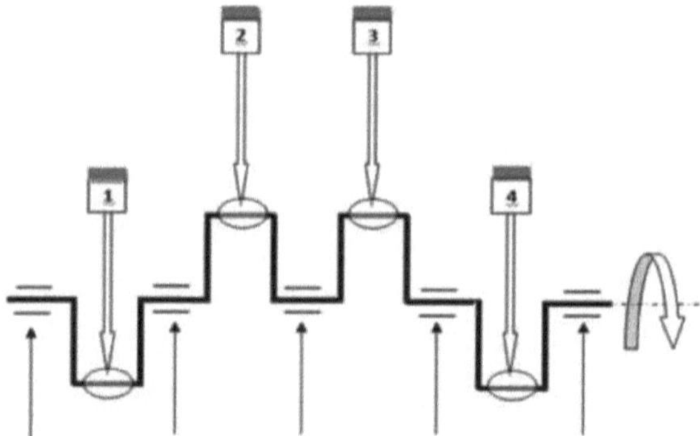

Trunnions: 5 crankshaft bearings in the engine block.

Figure 5: Crankshaft
Source: SILVA, 2015

In a four-stroke engine, the explosion is observed at one of these times. This is when the maximum torque is obtained, i.e. when the lever arm between the trunnion and the journal is greatest. For this reason, an internal combustion engine with a very large trunnion to journal distance cannot reach high revs (above 6,000 RPM). On the other hand, an engine with a smaller trunnion to journal distance can reach higher revs (up to 8,000 RPM), and consequently produces more power and less torque (RIBAS, 2003).

2.5.5 Torque and Power

According to Gillespie (1992), torque and power can be related as a function of speed. In the case of Otto cycle engines, the torque peak is generally at the middle of the speed range, while in Diesel cycle engines the torque peak is at low speeds, and the curve is more rectilinear compared to Otto cycle engines.

Figures 6 and 7 show examples of torque curves for Otto and Diesel cycle engines.

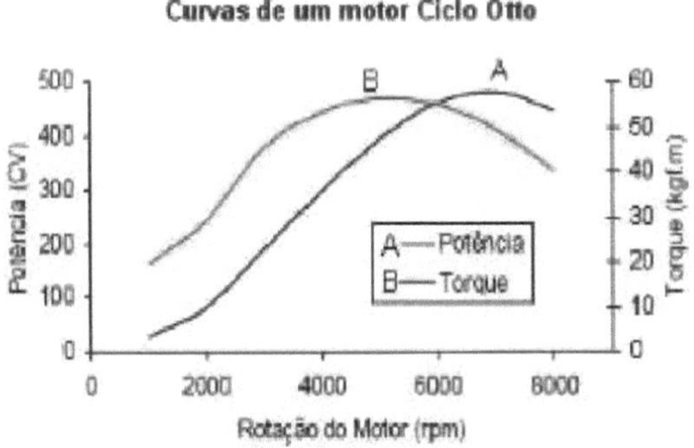

Figure 6: Curves of an Otto Cycle Engine
Source: *(GILLESPIE, 1992)*

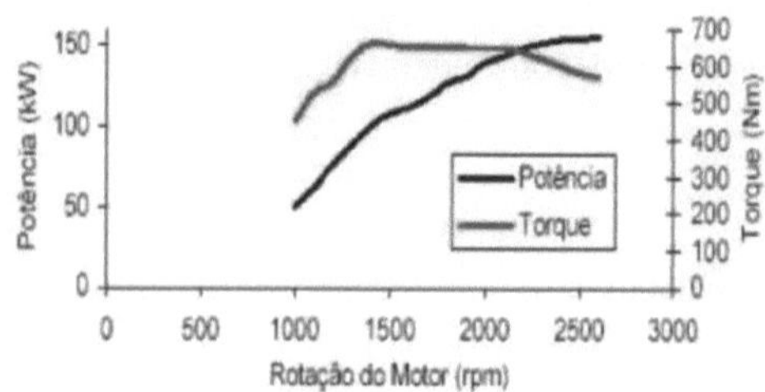

Figure 7: Diesel engine curves
Source: *(GILLESPIE, 1992)*

Torque values are measured on dynamometers. Using these values, power can be obtained from Equation (6).

$$P = \tau.\,\omega \tag{6}$$

Where:

(P) is power, which in SI is expressed in watts 1 W=1 Nm.s⁻¹ .

(T) is the torque.

(Ω) is the rotation.

Carvalho (2011), carrying out various tests on an Otto cycle internal combustion engine, observed that in the performance test, type C gasoline had a loss of torque and power compared to hydrated alcohol.

2.5.6 Fuel efficiency

Each engine has a different power output and consequently a different fuel consumption, due to the fact that power and other variables are linked to fuel consumption. To calculate fuel efficiency, the mileage per liter is used, where in SI it is given by Km l (SILVA, 2016), as shown in Equation (7)

$$km\,/\,1 = mileage\,/\,liter \tag{7}$$

According to Carvalho (2011) hydrated alcohol has a higher fuel consumption than type c gasoline, due to the fact that alcohol has a higher anti-detonation index and a faster burning speed.

2.6 EMISSIONS FROM INTERNAL COMBUSTION ENGINES

The generation of gas emissions is due to the combustion of the air/fuel mixture. The combustion or burning of fuel in internal combustion engines takes place in a closed enclosure called a combustion chamber, made up of three elements: fuel, oxygen and heat, which in the Otto cycle is generated by the spark and in the diesel cycle by compression (MANAVELLA, 2009).

Figure 8 shows the three elements of combustion.

Figure 8: Elements of Combustion
Source: MANAVELLA, 2009

This combustion or burning produces gases that are expelled through the exhaust pipe. Some of these gases are not harmful, such as O_2 (oxygen) and CO_2 (carbon dioxide), however the presence of CO (carbon monoxide) and HC (hydrocarbons) are extremely harmful to health (MARTINS, 2013).

After years of study and research into the formation of pollutants from internal combustion engines, a number of technologies have been developed to manage the control and monitoring of emissions and the treatment of exhaust gases. As a result, vehicle emissions levels can be controlled and regulated. As a result, certain maximum emission levels have been created in order to reduce environmental pollution. As a result, several countries around the world have regulated emission limits (HEYWOOD, 1988).

In Brazil, the body responsible for deliberating and establishing legal limits for vehicle emissions is the National Environment Council (CONAMA).

To implement the resolutions, CONAMA created two air pollution control parameters, one for motor vehicles, PROCONVE, which is used for light vehicles, and PROMOT, which is used for motorcycles and similar vehicles, setting deadlines, maximum emission limits and establishing technological requirements for domestic and imported motor vehicles (CETESB, 2011).

According to Cetesb (2011), because emission limits are getting lower and stricter, vehicle and fuel manufacturers and component suppliers will join forces with the academic and scientific community to find more efficient solutions to comply with legislation.

Legislation in Brazil and in several other countries has led to a significant reduction in emissions.

In order to comply with this legislation, a number of technologies have been developed to manage the control and monitoring of emissions and the treatment of exhaust gases (BOSCH, 2005).

One of the technologies that has become of great importance in complementing efforts to minimize emission levels is the catalyst. There are two types of catalyst, oxidizing catalysts for HC and CO, which in addition to being thermal reactors are filters for particulate materials, and basically reducers for NOx, and three-way catalysts for all three gases (Figure 9). Three-way catalysts basically consist of a ceramic substrate with an active coating incorporated, containing noble metals such as platinum, palladium and rhodium (MARTINS, 2013).

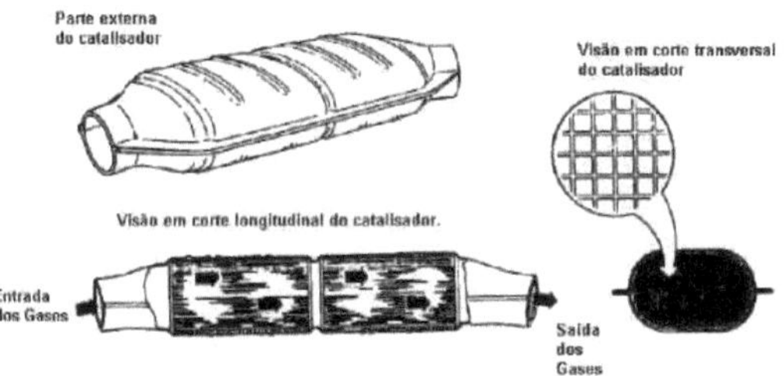

Figure 9: Three-way automotive catalytic converter
Source: (HEYWOOD, 1988)

After carrying out several analyses of pollutant gas emissions in internal combustion engines of various vehicles, Pozzagnolo (2013) reports that with technological advances in engines, there has been a great reduction in the pollutant gases emitted by vehicles using either hydrated alcohol or type c gasoline.

3 MATERIAL AND METHODS

This chapter presents the procedures used to carry out the research, including a description of the vehicles and equipment used in the tests and trials. In addition to the tests and trials, calculations will be carried out to obtain some of the data that will also be presented in this chapter.

3.1 CHARACTERISTICS OF THE VEHICLES AND FUELS USED IN THE TESTS AND TRIALS

3.1.1 S10 Rodeo Flexpower

The S10 flexpower rodeo is from the Chevrolet brand, which has a 4-cylinder in-line Otto cycle engine with electronic fuel injection, 8 valves, multi-point injection, 2,405 cm³ of displacement and is dual-fuel, making it possible for it to run on both ethanol and gasoline. The engine is located at the front of the truck and is well cooled, which is a good advantage, but it has the disadvantage that the truck is rear-wheel drive, which means that the engine loses power until it reaches the wheel. This is due to a complex driveshaft system, known as a cardan, which is used to transmit the force from the engine to the wheels, generating greater weight for the vehicle and reducing the interior space (ICARROS, 2012).

3.1.1.1 Petrol

The gasoline used in the tests and trials in this research is ordinary Brazilian gasoline, or type C, purchased from one of the service stations in the city of Foz do Iguaçu. This gasoline contains 25% anhydrous alcohol and 75% pure gasoline.

3.1.1.1 Ethanol

The ethanol used was the common type, hydrous ethanol, which is sold at different gas stations.

3.1.2 S10 Rodeio Diesel

The S10 rodeo diesel is from the Chevrolet brand and contains a diesel cycle engine with electronic injection, 4 cylinders in line, 8 valves, common rail diesel power, 2,799 cm³ of displacement. The S10 rodeo diesel has the same characteristics as the S10 rodeo flexpower, so it is known that the truck's engine is located at the front, generating good cooling, which is a good advantage, but it has the disadvantage that it is rear-wheel drive, which generates a loss of power from the engine to the wheel. This is due to a complex driveshaft system known as a cardan. The cardan is used to transmit the force from the engine to the wheels, generating more weight for the vehicle and reducing the internal space (ICARROS, 2012).

The S10 rodeo diesel that was used in this research is repowered with a chip for greater performance. Because of this, the torque and power tests were carried out separately, with and without the chip, to see what difference it made.

3.1.2.1 Diesel

S10 diesel oil contains a maximum sulphur content of 10mg kg⁻¹ . It is usually found at any petrol station and is the most recommended by vehicle manufacturers.

3.2 PERFORMANCE TESTS

To carry out the performance tests there is a standard, NBR 1585, which in the case of this research was not followed due to time and resource issues, but the tests were carried out on equipment called a dynamometer, which is equipment capable of measuring the engine's torque and power, These measurements are obtained through a graph that is generated by a device, which receives information from a roller that has a known weight where the vehicle with its drive wheels are placed, after which the vehicle is fully accelerated in a certain gear, thus obtaining the torque and power at the wheel. The device that generates the graph of power and torque as a function of engine speed or speed generates nothing more than a 'photograph' of the engine under acceleration (SERVITEC, 2010).

The torque measured is the force of the engine itself, which in the case of the dynamometer is what the wheel delivers to the roller plus the losses, while the power is the work done by the engine in a unit of time, in practice it is the product of the force of the engine by its rotation in a given time. Note that the power generated in the graph is that generated by the vehicle's wheel. To obtain that of the engine, the losses of the transmission system known as the cardan must be taken into account (GILLESPIE, 1992).

The graph generated by the device is used for visual interpretation, which covers various factors, including the speeds at which the engine is strongest or weakest, the overall quality of the curve, such as the average power, the speeds at which the engine has the best performance, whether there are faults in the fuel supply or mechanical faults, among others (SERVITEC, 2010).

Some checks, such as oil and water levels, were carried out before the tests, which are essential for running on the dynamometer, mainly due to safety issues. The tests are carried out under conditions of high speeds, pressures and temperatures.

To carry out the tests, the vehicle's drive wheel was placed on top of the roller as shown in Figure 10.

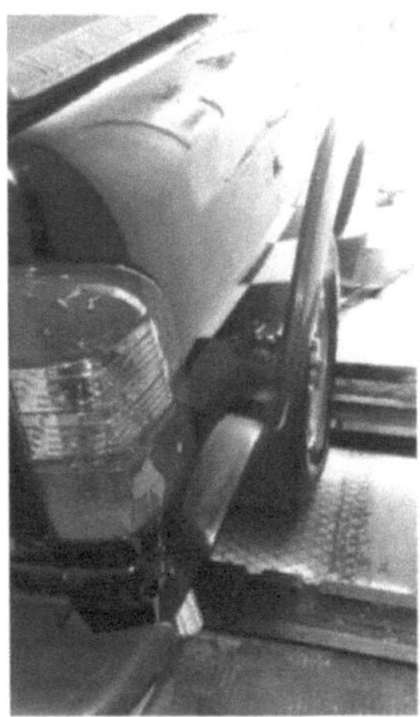

Figure 10: Dynamometer
Source: (Prepared by the author)

The ratchets were then fitted with a heavy-duty strap, so that the vehicle would not move during the test. The vehicle was then started and idled until it reached the working temperature, i.e. 90 degrees, after which the test was carried out.

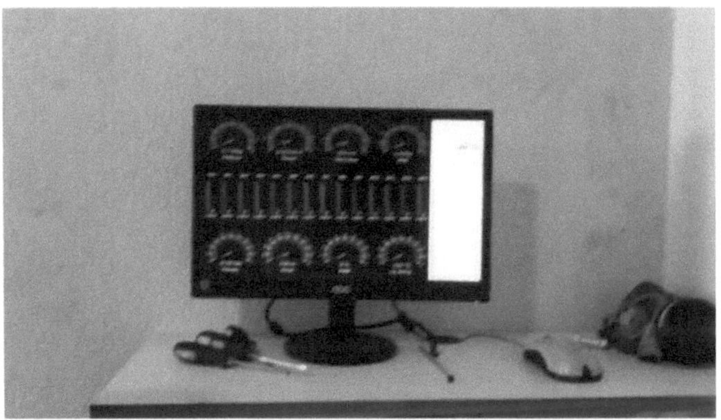

Figure 11: Dynamometer **Source:** (Prepared by the author)

3.3 POLLUTANT EMISSION TESTS

The pollutant emission tests were carried out in two ways, because pollutants are not measured in diesel cycle engines, but rather opacity. For both tests there is the NBR 12897 standard, which was followed, as both tests were carried out in suitable locations. For the S10 Rodeio flexpower truck, which uses an Otto cycle engine, a napro brand gas analyzer was used. A bosch opacimeter was used for the S10 Rodeo diesel.

3.3.1 Opacimeter

The opacimeter is a portable instrument made up of an optical bench, a probe (cable inserted into the exhaust) and a case with cables. It is used to measure the amount of particulate matter (black smoke) emitted by diesel vehicles. To report the result, the opacimeter uses the K-index measurement, which is the amount of smoke emitted per meter (BOSCH, 2005).

When the tests were carried out, the opacimeter was switched on and a self-test was carried out, which the device did on its own, after which it was placed in the vehicle (Figure 12).

Figure 12: Opacimeter
Source: (Prepared by the author)

The vehicle was then accelerated to its limit, generating a particulate material that the opacimeter measured in the form of a k index, the test was performed 10 times due to the variation between the results.

3.3.2 Gas Analyzer

The gas analyzer is a microprocessor-based instrument that, using a suction pump, sucks in gases produced by the vehicle, passing them through electrochemical cells that will give the concentration of the gases you want to analyze. Through this analysis, the types of gases produced by the vehicle and their concentrations can be obtained (MARTINS, 2013).

To carry out the tests, the gas analyzer was turned on and a self-check was made, which the device did on its own, after which it was placed in the vehicle (Figure 13) and the system was decontaminated.

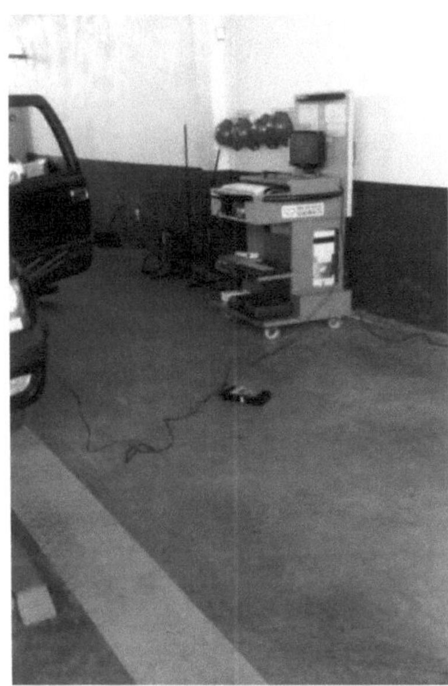

Figure 13: Gas Analyzer **Source:** (Prepared by the author)

Afterwards, a test was carried out in which the vehicle went through two stages, one in which it was accelerated (2500 rpm) and the other in which it was left idling, with the device measuring the gases generated by the vehicle in these two ranges.

3.4 FUEL EFFICIENCY TESTS

To carry out the fuel efficiency tests, there is a standard (NBR 7024), which in the case of this research was not followed for reasons of time and resources, but the tests were carried out using a 5-liter gallon, which was inserted in the two vehicles used, apart from this factor, the same route was stipulated for the 3 tests carried out, two of which were on the S10 flexpower (one using alcohol and the other petrol) and one on the S10 diesel.

Through these tests, data was collected such as the amount of mileage the vehicle covered and the fuel consumed in relation to the 5 liters initially inserted. After collecting this data, a calculation was made for each fuel used. After carrying out the tests, Equation (7) was used to determine the fuel consumption.

$$km / l = mileage / liter \qquad (7)$$

4 RESULTS AND DISCUSSION

This chapter presents the results of the tests carried out during the research, which will be discussed and compared with the findings of other authors. Figures and tables will be presented for a better understanding of the results. Due to the fact that some results are not very clear, tables have been created, the originals of which can be found in Appendix A.

4.1 PERFORMANCE TEST RESULTS

In the performance test, it was possible to analyze the results using the equations and images of the graphs generated by the dynamometer.

The equation is given by the device itself (Equation 8).

$$Fm = Fr + 15\%.Fr + 10HP \tag{8}$$

Figures 14, 15, 16 and 17 show the torque and power results of the vehicles using each of the fuels (alcohol, gasoline and diesel).

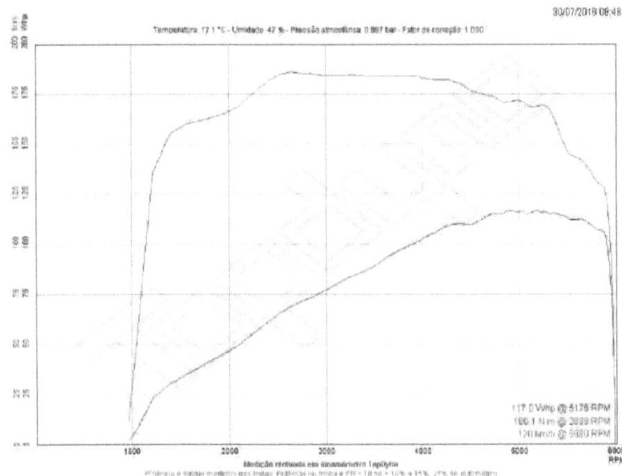

Figure 14: Alcohol torque and power test results
Source: (Prepared by the author)

Because the device reports the torque and power of the wheel and not the engine, especially in the alcohol and petrol tests, it was necessary to use Equation 8 to obtain the corresponding torque and power of the engine.

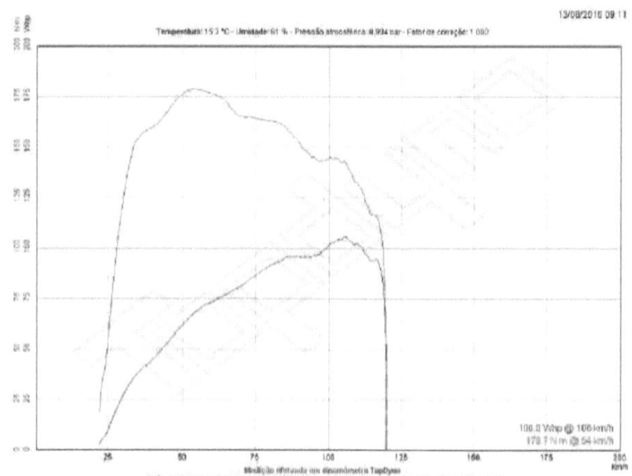

Figure 15: Gasoline torque and power test results **Source:** (Prepared by the author)

Figure 15 shows the results of the gasoline torque and power test, which was carried out on the s10 rodeo flexpower truck. A comparative analysis of the gasoline and alcohol curves shows that the gasoline torque curve decays more quickly than the alcohol torque curve.

Alcohol has a higher anti-knock index than gasoline, in which parameters such as the engine's compression ratio, ignition advance time and stoichiometric ratio can be varied, generating an increase in pressure inside the chamber, resulting in higher torque and power values (BOSCH, 2005).

Alcohol also has excellent octane qualities, providing a cleaner burn, generating good properties for use in spark ignition engines, better than gasoline (OWEN AND COLEY, 1995).

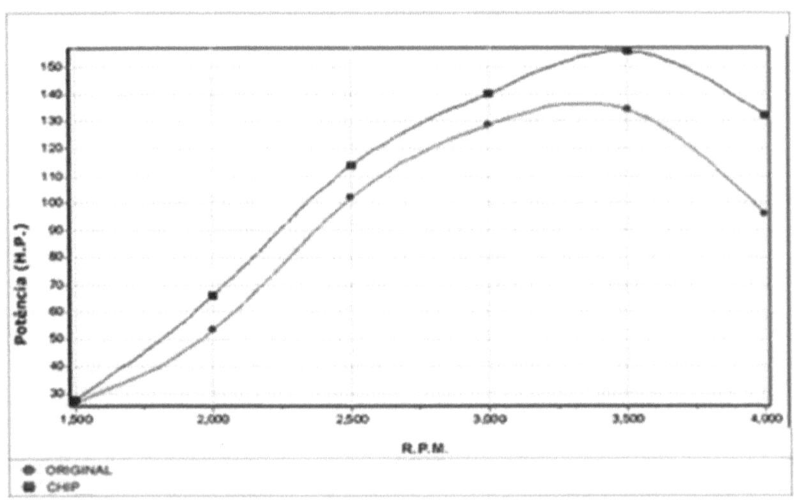

Figure 16: Diesel power test results

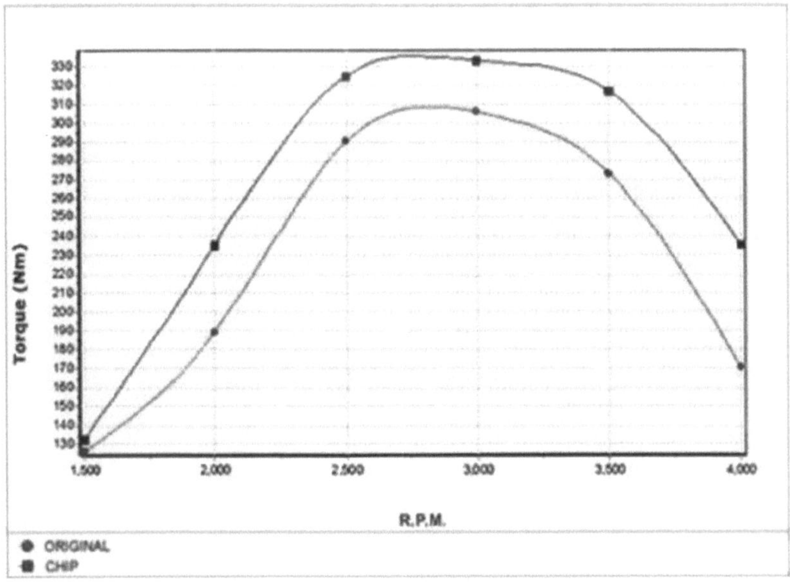

Figure 17: Diesel torque test results
Source: (Prepared by the author)

It can be seen that in the diesel tests (Figures 16 and 17) the device already reported the torque and the corresponding power of the engine and not the wheel, because the tests were carried out separately, which generated two curves for torque and two for power, one of them with a repower chip and the other original.

Table 1 shows the maximum torque and power values obtained from the graphs and equations for each fuel.

Table 1: Performance Test Results

Fuels	Torque	Power
Alcohol	214.015Nm	144.55HP
Petrol	205.505Nm	131.9HP
Diesel	305.8Nm	134.3HP

Source: (Prepared by the author)

These results corroborate those of Carvalho (2011), who found that in the performance test, type C gasoline had a loss of torque and power compared to hydrated alcohol, although the torque of diesel was higher than that of alcohol.

The results show that gasoline lost out compared to alcohol, due to the fact that alcohol has a higher anti-knock index than gasoline.

Compared to diesel, alcohol had a loss of torque, but maintained the highest power, due to the fact that the diesel cycle engine has ignition generated by compression between fuel and air, obtaining a higher compression ratio, which generates greater torque than Otto cycle engines, where ignition occurs by the spark, obtaining a lower compression ratio.

4.2   POLLUTANT EMISSIONS TEST RESULTS

In the pollutant emissions tests, two types of equipment are used: one for diesel (opacimeter), which measures the k-index, and the other for gasoline and alcohol (gas analyzer), which measures the types of gases.

Table 2 shows the test carried out on the opacimeter, which was used in the s10 rodeo diesel van.

Table 2: Opacimeter test results (diesel)

Number	K Index
K1	4.58 $l.m^{-1}$
K2	2.35 $l.m^{-1}$
K3	1.93 $l.m^{-1}$
K4	1.70 $l.m^{-1}$
K5	2.62 $l.m^{-1}$
K6	2.85 $l.m^{-1}$
K7	2.12 $l.m^{-1}$
K8	2.11 $l.m^{-1}$
K9	1.32 $l.m^{-1}$
K10	2.20 $l.m^{-1}$
Average	2,378 $l.m^{-1}$
Limit Value	2.10 $l.m^{-1}$

Source: (Prepared by the author)

The device reports the limit value specified by the manufacturer for each vehicle, in the case of the s10 rodeo diesel it is 2.10 $l.m^{-1}$. The test was carried out 10 times due to the variation, some of which were quite significant and others not (Table1).

According to the test carried out, the truck failed due to its high rate of smoke per meter.

Diesel may have high torque, but it should only be used for large or heavy vehicles, heavy machinery, agricultural machinery and so on, because even with advances in technology, diesel engines are very polluting.

Tables 3 and 4 show the results of the gasoline and alcohol tests, which used the gas analyzer to carry out the tests on the s10 rodeo flexpower van.

Table 3: Gas Analyzer Test Results (Gasoline)

Limits					
Low gear	2500 RPM	COc (%vol)	Factor. Dilution	HCc(ppm vol)	
600/1200 RPM	2500/2700 RPM	0,30	Maximum 2.50	100	
Petrol	Measured Values				
1ESC	RPM	COc (%vol)	Factor Dil.	HCc (ppm)	CO (%vol)	CO2 (%vol)	HC (ppm)
March	540	00,1	1,00	0	0,01	15,0	0
Slow							
2500 RPM	2550	00,1	0,99	1	0,01	15,1	1

Source: (Prepared by the author)

Table 4: Gas Analyzer Test Results (Alcohol)

Limits					
Low gear	2500 RPM	COc (%vol)	Factor. Dilution	HCc(ppm vol)	
600/1200 RPM	2500/2700 RPM	0,50	Maximum 2.50	250	
Alcohol	Measured Values				
1ESC	RPM	COc (%vol)	Factor Dil.	HCc (ppm)	CO (%vol)	CO2 (%vol)	HC (ppm)
March	800	00,1	0,98	0	0,01	15,3	0
Slow							
2500 RPM	2590	00,1	0,98	0	0,01	15,3	0

Source: (Prepared by the author)

In the gasoline test (Table 3) it can be seen that the vehicle failed, but not because of the polluting gases, but because of a variation in the idling speed that occurred due to a technical problem with the truck, which was fixed later; for the alcohol test the vehicle passed without any variation.

It can be seen that gasoline or alcohol did not have a high rate of polluting gases, which may have contributed to the year of manufacture (2011). Collaborating with the results of Pozzagnolo (2013) who, through various tests carried out, was able to arrive at results that proved that with technological advances in engines, there was a great reduction in the polluting gases emitted by vehicles using either hydrated alcohol or type c gasoline.

By comparing the test results for gasoline and alcohol, it can be seen that there is not much difference between the two. Compared to diesel, gasoline and alcohol are less polluting to the environment. However, the fuel that would have the least environmental degradation in terms of pollutant emissions would be alcohol, as it has the lowest values.

4.3 FUEL EFFICIENCY TEST RESULTS

In the fuel efficiency tests, alcohol covered 32 km, gasoline 50 km and diesel 45 km in the city with 5 liters of fuel. Using Equation 7 for each fuel.

Table 5 shows the results obtained using the equations.

Table 5: Fuel Efficiency Test Results

Fuel	Media
Alcohol	$6.4\ Km.l^{-1}$
Petrol	$10\ Km.l^{-1}$
Diesel	$9\ Km.l^{-1}$

Source: (Prepared by the author)

A comparison of the results shows that alcohol had the worst fuel efficiency, confirming the report by Carvalho (2011), who states that hydrated alcohol has a higher fuel consumption compared to type c gasoline, due to the fact that alcohol has a higher anti-detonation index and a faster burning speed than gasoline, but it is worth remembering that gasoline and diesel are derived from petroleum, which is not renewable, while alcohol is a renewable fuel, as it is derived from sugar cane.

In addition to this issue, alcohol performs better than gasoline in terms of both torque and power, and only loses out to diesel in terms of torque, as it maintains a better result in terms of power.

In terms of environmental degradation through pollutant emissions, alcohol has the lowest emissions, compared to gasoline, which had the lowest emissions, and diesel, which had the highest emissions.

A comparative analysis of all the results shows that alcohol is an excellent fuel, with excellent octane qualities, providing a cleaner burn.

Alcohol only needs to be improved in terms of performance, which is the worst, but given the scarcity of oil and the environmental degradation caused by emissions, alcohol is a good alternative and has been on the market for some time.

5 FINAL CONSIDERATIONS

The results of this research showed that of the two vans used, the S10 rodeo flexpower using alcohol is a viable solution to issues such as environmental pollution and the scarcity of oil, as it obtained the best result in the pollutant emission test and is a renewable fuel. In other tests, alcohol obtained excellent torque and power values compared to the other fuels used in this research, losing out only in the performance test, in which gasoline obtained the best result.

The diesel-powered S10 Rodeo had great torque and reasonable fuel efficiency, but on the downside it scored very poorly in the pollutant emissions test, making it less viable for passenger cars.

Given these considerations, it is clear that there are still great opportunities to improve the results of the studies, especially in relation to the emergence of new fuels and yields.

The different types of fuel used in internal combustion engines show differences in performance results, fuel efficiency and emissions. However, along with this research, it is important to study new fuel sources for internal combustion vehicles, whether hybrid or electric engine technologies. Continued research and technological innovation in internal combustion engines should focus on improvements in fuel efficiency, reductions in emissions and the search for new types of fuel that are renewable, bring better performance results and contribute to the environment and sustainability.

REFERENCES

ABIANEH, O.S. *et al,* **Combustion Development of a Bi-Fuel Engine**.
International Journal of Automotive Technology, v. 10, no. 1, pp. 17-25 (2008).

ALBAHRI, T. A. *et al,* **Analysis of Quality of the Petroleum Fuels**. Energy &
Fuels, 2003.

ALEME, H. G., **Determination of physical-chemical parameters of diesel oil from
distillation curves using chemometric techniques.** Doctoral thesis. Department of
Chemistry, Federal University of Minas Gerais, 2011, 149 p.

ANFAVEA. Yearbook of the Brazilian Automobile Industry. 2015.

BOSCH, Robert. Gmbh, *Automotive Handbook.* Verlag, Germany, 1993, 852p.

BOSCH, Robert Gmbh., **Handbook of Automotive Technology**. 25.ed. São Paulo:
Edgar Blucher, 2005. 1229 p.

CARVALHO. M. A. S., **Evaluation of an otto cycle internal combustion engine
using different types of fuel**. Salvador: UFBA, 2011 (Master's thesis).

CETESB. Information on emissions regulations. 2011.

CHO, H. M.; HE, B., **Combustion and Emission Characteristics of a Lean Burn
Natural Gas Engine. International Journal of Automotive Technology**, Vol. 9,
no. 4, pp. 415-422 (2008).

FERNANDES, R. K. M. *et al.,* **Biodiesel from waste frying oil: an energy
alternative and socio-environmental development**. In: XXVIII NATIONAL
PRODUCTION ENGINEERING MEETING, 2008, Rio de Janeiro.

FERREIRA, S. M., **Relation between the Real and Apparent Thickness of
Gasoline with Ethanol and Pure Gasoline in Laboratory Columns**. PhD thesis
presented at USP, 2003.

GILLESPIE, Thomas. D., **Fundamentals of Vehicle Dynamics.** United States of
America: Society of Automotive Engineers, INC, 1992.

KNOTHE, G.; *et al.* Biodiesel: the use of vegetable oils and their derivatives as
alternative diesel fuels. *ACS Symp Ser,* v.666, p.172-208, 1997.

HALLIDAY, David *et al.,* **Fundamentals of Physics.** 8. ed. Rio de Janeiro: LTC,
2008.

HEYWOOD, J. B., **Internal Combustion Engine Fundamentals**.
McGraw-Hill
International ed, 1988, 930p.

ICARROS - Icarros Technical Sheet. Available at:
< http://www.icarros.com.br/chevrolet/s10-cabine-dupla/2011/ficha- tecnica/7295>
Accessed: Aug. 6, 2016.

ICARROS - Icarros Technical Sheet. Available at:
<http://www.icarros.com.br/chevrolet/s10-cabine-dupla/2011/ficha- tecnica/7296>
Accessed on: Aug. 6, 2016.

JEWETT, Junior John W.; SERWAY, Raymond A., **Physics for Scientists and
Engineers.** 1. ed. São Paulo: CENGAGE LEARNING, 2012.

JR, L. C. M., **Internal Combustion Engines - Basic Concepts.** Rio Grande do Sul.
Unijuí, 2008.

MANAVELLA, Humberto José, **Combustion and Emissions.** 1. ed. São Paulo: HM
Autotrônica, 2009.

MARTINS, Jorge, **Internal Combustion Engines.** 4.ed. Porto: Publindustria -
Edições Técnicas, 2013. 480 p.

MOTHÉ, C. G. *et al.,* Optimization of biodiesel production from castor oil. *Revista
Analítica,* v.1, n.19, p.40-43, 2005.

OICA. The International Organization of Motor Vehicle Manufacturers. 2011.

OWEN, K.; COLEY, T., **Automotive Fuels Reference Book**. 2nd ed.
SAE, 1995.

POMPELLI, M. F. et al, World energy crisis and Brazil's role in the biofuel problem.
Revista Agronomia Colombiana, Colombia, p. 231-240, February 2011.

PETROBRÁS - Petróleo Brasileiro S/A. Available at:
<http://www.petrobras.com.br>. Accessed on: May 13, 2016.

POZZAGNOLO. M., **Analysis of gas emissions in otto cycle motor vehicles**.
Lajeado: UNIVATES, 2013. (Final course work).

RAMALHO, Junior Francisco; *et al.,* **The Fundamentals of Physics.** 9. ed. São
Paulo: MODERNA, 2007.

RIBAS, R. B., **Mecânica de Automóveis Motores de Combustão Interna - Álcool**

e Gasolina. Porto Alegre, 2003.

SERVITEC -. Servitec Dynamometer Available at: <http://www.servitecdinamometro.com.br>. Accessed on: May 13, 2016.

Silva, Marcos R. **Combustion engine: how it works - part 2.** Valinhos, 2014.

SILVA, Marcos Noé Pedro Da. "Calculating the Fuel Consumption of an Automobile"; *Brasil Escola.* Available at <http://brasilescola.uol.com.br/matematica/calculo-consumo- combustivel-um-automovel.htm>. Accessed on May 29, 2016

SOCCOL, C. R. *et al.* Brazilian biofuel program: an overview. *Journal of Scientific and Industrial Research,* v. 64, p. 897- 904, 2005.

Tecnomotor - Products. Available at: <http://www.tecnomotor.com.br/novo/index.php/produtos/2015-04-02-20- 05-00/tm-131-analisador-de-gases>. Accessed on: May 19, 2016.

TILLMANN, Carlos Antonio da Costa, **Internal combustion engines and their systems**. Pelotas, 2013.

TORRES, Marcelo A. S., **COMBUSTION ENGINES.** GUARATINGUETÁ, 2010.

VALENTE, O.S., **Performance and emissions of an engine-generator operating with biodiesel.** Master's dissertation in Mechanical Engineering, Catholic University of Minas Gerais, 2007, 161p.

VARELLA, Carlos Alberto Alves, **HISTORY AND DEVELOPMENT OF INTERNAL COMBUSTION ENGINES.** RIO DE JANEIRO, 2010.

APPENDIX A - Pollutant emission test results

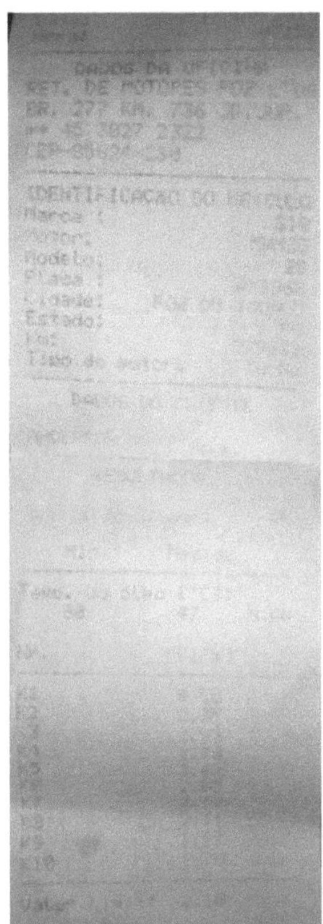

Figure 18: Opacimeter test result
Source: (Prepared by the author)

PC - MULTIGAS

OTTO CYCLE VEHICLE INSPECTION

INSPECTION REPORT

FAILED

Customer NameANDERSON BASSETORenavam		n°: 256799725
License plate: ATF-4443		Year of Manufacture: 2010 / 2011
Brand: GM	Model: S10 RODEIO	DCFuel: Petrol
Date: 27/09/2016	Start time: 08:07:11 End time	: 08:12:09
Category:		Oil Temperature (Engine)90 (°C)

Defects Visual Inspection

No Defects Found

Rotation check:

Off-range idle speed: 540 RPM (limits 600 to 1200)

Gas Measurement

Limits				
Low gear	2500 RPM	COc (%vol)	Dilution Factor	HCc(ppm vol)
600/1200 RPM	2300/2700 RPM	0,30	maximum 2.50	100

Petrol	Measured Values						
1ESC	RPM	COc (%vol)	Dil factor.	HCc(ppm)	CO (%vol)	CO2 (%vol)	HC(ppm)
Low gear	540	0,01	1,00	0	0,01	15,0	0
2500 RPM	2550	0,01	0,99	1	0,01	15,1	1

Figure 19: Gas Analyzer Test Result (Gasoline)
 Source: (Prepared by the author)

PC - MULTIGAS

OTTO CYCLE VEHICLE INSPECTION

INSPECTION REPORT

APPROVED

Client Name:ANDERSON BASSETORenavam	n°: 256799725
License plate: ATF-4443	Year of Manufacture: 2010 / 2011
Brand: GM	Model: S10 RODEIO D Fuel: Alcohol
Date: 06/10/2016	Start time: 07:44:48 End time : 07:50:29
Category:	Oil Temperature (Engine):91 (°C)

Gas Measurement

Limits				
Low gear	2500 RPM	COc (%vol)	Dilution Factor	HCc(ppm vol)
600/1200 RPM	2300/2700 RPM	0,50	maximum 2.50	250

Alcohol	Measured Values						
1ESC	RPM	COc (%vol)	Dil factor.	HCcfppm)	CO(%vol)	CO2 (%vol)	HC(ppm)
Low gear	800	0,01	0,98	0	0,01	15,3	0
2500 RPM	2590	0,01	0,98	0	0,01	15,3	0

Figure 20: Gas Analyzer Test Result (Alcohol)
Source: (Prepared by the author)

APPENDIX B - Equipment Calibration and Verification Protocol

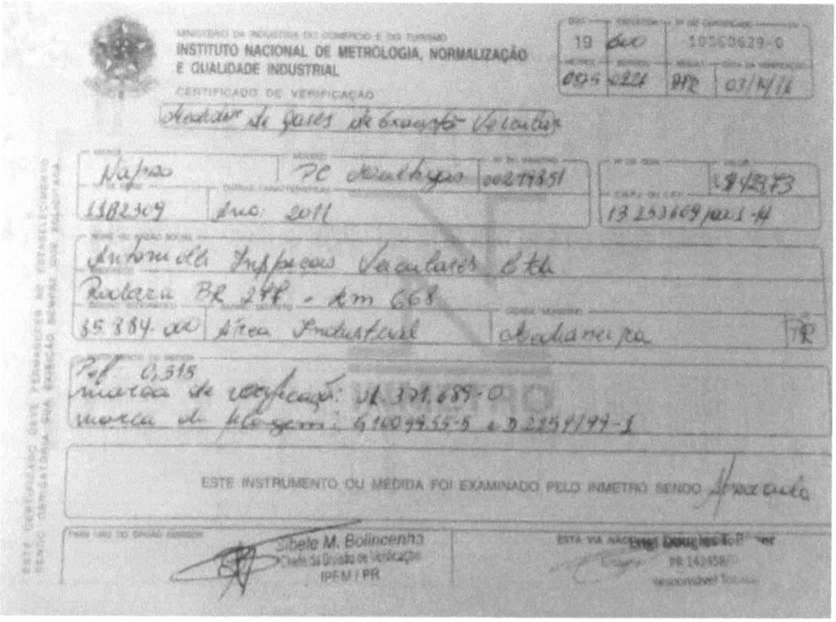

INSTITUTO NACIONAL DE METROLOGIA, NORMALIZAÇÃO E QUALIDADE INDUSTRIAL

CERTIFICADO DE VERIFICAÇÃO

ESTE INSTRUMENTO OU MEDIDA FOI EXAMINADO PELO INMETRO SENDO

Sibele M. Bolincenha
Chefe da Divisão de Verificação
IPEM / PR

Figure 21: Gas Analyzer Verification Protocol **Source:** (Prepared by the author)

TECMETRON ▪ **Autorizada Bosch**

de: IVALDO SIPRIANO DE OLIVEIRA

EQUIPAMENTOS PARA OFICINAS

RUA FRANCISCO DEROSSO, 1163 - LOJA 04 - FONE/FAX (41) 3275-4352
CEP 81710-000 - CURITIBA - PARANÁ

Serviços Técnicos Eletrônicos e Eletro Mecânicos em Equipamentos de Oficinas e Bancada de Teste para bombas injetoras.

Nome: Retificadora de motores foz LTDA

End: BR 277 KM 728, 250 terreo

Cidade: Foz do Iguaçu – PR

CNPJ: 00.170.924/0001-36

IE: 422.117.628-1

Protocolo de calibração e verificação

Certificado de calibração Nº 02/2016 Aparelho série Nº 320.016.624

Nº do opacimetro: 0684.103.119   Nº de série: 320.016.624

DF: 03/2002

Meios de teste: especificado

Pino de calibração padrão Bosch   Nº de série: 1688.130.220

PS	Passo do teste	Nominal	Min	Valor medido (VM)	Máx	Função/Valor medido SIM cumprida NÃO
1	Precisão de indicação		45%		55%	
1,1	Precisão de indicação (Turvação)	Valor pino de calibração	1,40 m-1 do VM		1,80 m-1 do VM	

X _(assinatura)_
Ivaldo S. de Oliveira

84.814.987/0001-78

IVALDO SIPRIANO DE OLIVEIRA
TECMETRON ME

RUA FRANCISCO DEROSSO, 1163 LJ. 04
XAXIM - CEP 81.710-000
CURITIBA - PR

Verificação: APROVADA

Data do teste: 02/09/2016   Próxima aferição: 02/09/2017

Nome do examinador: Ivaldo S. de Oliveira

Figure 22: Opacimeter Calibration and Verification Protocol
Source: (Prepared by the author)

yes
I want morebooks!

Buy your books fast and straightforward online - at one of world's fastest growing online book stores! Environmentally sound due to Print-on-Demand technologies.

Buy your books online at
www.morebooks.shop

Kaufen Sie Ihre Bücher schnell und unkompliziert online – auf einer der am schnellsten wachsenden Buchhandelsplattformen weltweit! Dank Print-On-Demand umwelt- und ressourcenschonend produziert.

Bücher schneller online kaufen
www.morebooks.shop

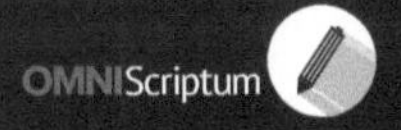

Printed by Books on Demand GmbH, Norderstedt / Germany